Chapter 1
ADMIN

Section 109
INSPECTIONS

Chapter 2
DEFINITIONS

Chapter 3
COMPL METHODS

Chapter 3
COMPL METHODS

Chapter 4
REPAIRS

Chapter 5
PRESCRIP
COMPL METH

Chapter 5
PRESCRIP
COMPL METH

Chapter 6
CLASS OF WORK

Chapter 7
ALTER - LEVEL 1

Chapter 7
ALTER - LEVEL 1

Chapter 8
ALTER - LEVEL 2

Tables 1011.5-1011.7
HAZARD CATEG'S
& CLASSIF

Chapter 11
ADDITIONS

Chapter 11
ADDITIONS

Chapter 12
HISTORIC BLDGS

Chapter 13
PERFORM
COMPL METH

Chapter 14
RELOC BLDGS

Chapter 15
CONSTRCT
SAFEGRDS

Chapter 16
REFERENCED STDS

Chapter A1
UNREINFRCD
MASONRY

Table A108.1(1)
STRENGTH VALS
EXIST'G MATLS

Table A108.1(1)
STRENGTH VALS
EXIST'G MATLS

Chapter A2
REINF CONC/MAS
W/ FLEX DIAPH

Chapter A2
REINF CONC/MAS
W/ FLEX DIAPH

Chapter A3
CRIPPLE/SILL ANCHOR

Chapter A3
CRIPPLE/SILL ANCHOR

Table A304.3.1
SILL PLT &
CRIPPLE BRACING

Table A304.3.1
SILL PLT &
CRIPPLE BRACING

Chapter A5
REFER STDS - APP A

Appendix B
APP B -
ACCESSIBILITY

Appendix B
APP B - ACCESSIBILITY

Appendix C
APP C - WIND RETROFIT

Appendix C
APP C - WIND RETROFIT

Appendix D
BOARD APPEALS

Appendix D
BOARD APPEALS

Appendix E
TEMP
EMERG USES

Resource A
ARCHAIC MATLS

Resource A
ARCHAIC MATLS

Table of Contents
RESOURCE A -
TOC

Table of Contents
RESOURCE A -
TOC

Appendix Sect. I
PERF - WALLS

Appendix Sect. I
PERF - WALLS

Appendix Sect. II
PERF - COLUMNS

Appendix Sect. II
PERF - COLUMNS

Appendix Sect. III
PERF - FLR/CLG

Appendix Sect. IV
PERF - BEAMS

Appendix Sect. IV
PERF - BEAMS

Appendix Sect. V
PERF - DOORS

Appendix Sect. V
PERF - DOORS